On the Go
Digging Machines in Action

David and Penny Glover

PowerKiDS press.

New York

Published in 2008 by The Rosen Publishing Group, Inc.
29 East 21st Street, New York, NY 10010

First Edition

Editor: Camilla Lloyd
Editorial Assistant: Katie Powell
Designer: Elaine Wilkinson
Picture Researcher: Kathy Lockley

Picture Credits:
The author and publisher would like to thank the following for allowing these pictures to be reproduced in this publication:
Cover: JCB and Volvo (inset); JCB: 2, 5, 6, 7, 11t, 13, 14, 16, 17, 20; Terex GmbH: 19; Volvo: 1, 4, 9, 18; Gunter Marx/Alamy: 10, AdrianSherrett/Alamy: 15; CP Stock/Constructionphotography.com: 8, 12; Charles O' Rear/Corbis: 11b, Michael Pole/Corbis: 21, Corbis: 22.
With special thanks to JCB, Terex GmbH and Volvo.

Library of Congress Cataloging-in-Publication Data

Glover, David, 1953 Sept. 4-
    Digging machines in action / David and Penny Glover.
        p. cm. — (On the go)
    Includes index.
    ISBN 978-1-4042-4307-1 (library binding)
    1. Excavating machinery—Juvenile literature. I. Glover, Penny. II. Title. III. Series.

    TA732.G585 2008
    621.8'65—dc22

                                2007032268

Manufactured in China

Contents

What are excavators?

Excavators are big digging machines. Excavators dig deep holes and long **ditches**. This excavator is digging a ditch for a new pipe.

excavator

ditch

4

excavator arm

trash

dumpster

Sometimes excavators pick things up. This excavator is putting trash in a dumpster.

Excavator quiz

Why is the excavator digging a ditch?

Excavator parts

cab

boom

bucket

The excavator driver sits in the **cab**. This is high up, so he can see all around. The excavator's arm is called a **boom**. It swings and bends to move the **bucket**.

teeth

The bucket is on the end of
the boom. It scoops up soil and rocks.
The bucket has sharp teeth to cut
through the ground.

Excavator quiz
Where is the excavator's bucket?

How does it work?

joystick

The driver uses **levers** called **joysticks** to make the excavator work.

The driver can turn the boom and the bucket to pick up a **load**.

joint

The boom bends
to move the load
up and down.
It has a
special **joint**,
like an elbow.
The bucket tilts
to empty its load.

Excavator quiz

**How does the driver
work the excavator?**

Tracks and wheels

tire

wheel

Some excavators move around on wheels. A rod called an **axle** attaches the wheels to the excavator's body. The wheels have thick, rubber tires to grip the ground.

tracks

links

Some excavators move around on **tracks**. Each track is a loop of metal links. Tracks help the excavator to grip the ground and stop it from slipping.

Excavator quiz
How are wheels attached to an excavator's body?

What makes it go?

fuel tank

The excavator's **engine** makes it go. The engine turns the wheels or tracks to drive the excavator along.

The engine runs on **diesel fuel**. The driver fills up the tank to make the engine work.

The engine powers the boom and bucket, too. The joint bends and special **pistons** make the parts move.

joint

piston

Excavator quiz
What kind of fuel does an excavator use?

Special jobs

Excavators move rocks and rubble to build new roads. This excavator has a **drill** to make holes for posts or to plant a tree.

drill

The junkyard excavator has a **grapple** instead of a bucket. The grapple grips like a hand to pick up metal.

grapple

Excavator quiz
Where can an excavator use a grapple?

Mini-excavators

A big excavator cannot fit inside a house, but a mini-excavator can. A mini-excavator works just like a big excavator, but it has a smaller bucket and boom.

It's hard work to dig a pond with a spade. A mini-excavator soon gets the job done.

Excavator quiz

What can a mini-excavator do?

Giant excavators

Giant excavators work in **quarries**, digging up rocks. The excavator breaks the rocks with its bucket. Its teeth are bigger and stronger than a dinosaur's.

The RH400 is the biggest excavator in the world. The RH400 is so big, it has beds and a bathroom inside.

Excavator quiz

How does a giant excavator break up rocks?

Digging safely

Excavators are very powerful machines. You must never go near an excavator when it is working. A safety fence keeps people away from danger.

safety fence

Excavators make lots of noise. Drivers wear ear protectors to protect their hearing. They always wear hard hats, just in case a loose stone hits their head.

Excavator quiz
Why does the driver wear a hard hat?

Old Excavators

steam excavator

The first excavators used steam to move. Before excavators were invented, people had to dig everything with spades. This **steam excavator** loads rock onto the train.

Excavator words

axle
The rod through the center of a wheel.

boom
The excavator's arm.

bucket
The heavy scoop on the end of the boom. The bucket has sharp teeth to cut into soil or rock.

cab
The part of the excavator in which the driver sits.

diesel
The fuel an excavator engine uses to make it go.

ditch
A long, narrow hole in the ground.

drill
A tool that turns to make a hole.

engine
The part of the excavator that makes it work.

fuel
Something that burns inside an engine to make it work.

grapple
A tool made of large spikes that can pick up pieces of metal.

joint
The part of the excavator's boom that bends, so that it can move the load up and down.

joystick
A lever that works the boom or bucket.

lever
A control like a stick that you move to make something work.

load
Something you lift or carry.

piston
The rod attached to the boom's joint. It makes the boom and the bucket move.

quarry
A place where rocks are dug up from the ground.

steam excavator
An old-fashioned excavator powered by steam.

track
The metal links that some excavators move around on instead of wheels.

Quiz answers

Page 5 For a new pipe.

Page 7 On the end of the boom.

Page 9 With joysticks.

Page 11 With axles.

Page 13 Diesel fuel.

Page 15 In a junkyard.

Page 17 Fit inside a house and dig a pond.

Page 19 With its bucket.

Page 21 In case a loose stone hits his head.

Index

Web Sites
Due to the changing nature of Internet links, PowerKids Press has developed an online list of Web sites related to the subject of this book. This site is regularly updated. Please use this link to access this list:
www.powerkidslinks.com/otg/digmach